Physical Science

Strategies for English Language Learners

CONTENTS

Strategies for Teaching
ENGLISH LANGUAGE LEARNERS

by Dr. Robin Scarcella
University of California at Irvine

Introduction

Teaching science to English Language Learners (ELLs) is both a challenge and an opportunity. It is a challenge because when students reach the upper grades, they face a demanding science curriculum and require advanced academic literacy to access this curriculum. It is an opportunity because science offers direct, interesting, hands-on, and minds-on experiences to students. These experiences will enable students to learn about the world, solve problems, and answer questions as they develop scientific and academic literacy.

Fortunately, a variety of motivating and effective means are available to teach reading in science classrooms. These means not only help your ELL students but also many of your monolingual English-speaking students. You can incorporate a deliberate focus on reading comprehension using a combination of reading comprehension strategies and language instruction.

The following pages describe effective practices that will help you meet the instructional challenges of teaching science to English language learners. They also show you how *Holt Science Spectrum* provides resources and strategies to support your instruction.

Improving Achievement Through Differentiated Instruction

▶ Why Is It Important?

English language learners have particular difficulty reaching challenging science standards. Their understanding of science standards generally lags far behind that of their peers whose native language is English. If left unaddressed, the poor ability of English language learners to reach basic physical science standards can have serious repercussions. It might even prevent them from later enrolling in courses required for college admission. Providing lessons that your students can understand and engage in will encourage students to participate in your lessons and will help them acquire key concepts and develop academic literacy.

▶ Strategies and Resources

Although students enter your classroom with a range of skills, all of them should be discussing the same topics, participating in similar activities, and completing similar assignments. You can help students improve their understanding of science concepts by providing them with reading and study tools and with additional instruction before school, after school, and during class.

To help you in these efforts, the *Holt Science Spectrum* resources are designed to teach English language learners difficult science concepts. These materials provide systematic suppport and resources for intervention in delivering a content-rich curriculum to English language learners. The Chapter Planning guide lists many examples of the resources at your disposal.

All the information you need to differentiate instruction and ensure that your students are reaching the science standards is provided in the **Chapter Planning Guide** in the **Teacher's Edition,** and in the **Lesson Plans** on the One-Stop Planner.

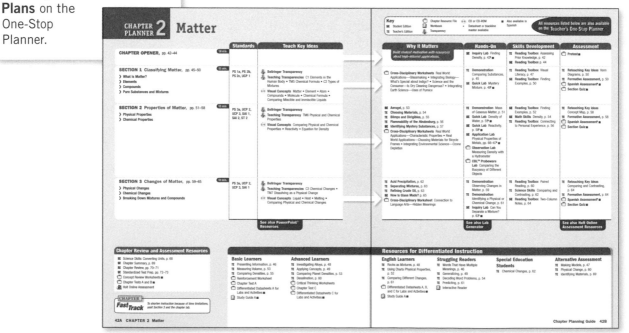

Interactive Reader

The **Interactive Reader** provides a simplified version of the **Student Edition** that focuses on key concepts and includes reading strategies and directed reading questions.

Interactive Reader

Differentiated Instruction

Struggling Readers

Using Headings Remind students that sec___ headings usually include important inform___ tion that will be explained in the section. H___ students copy each heading under the main heading *Energy and Reactions* on a sheet of paper. Point out that they now have a summ___ of the important points about energy and r___ tions. Have students scan the text under e___ B-head and find a sentence that gives m___ information about its statement. Have t___ copy this statement onto their paper ___ appropriate heading. **LS** Logical/V___

Throughout the **Teacher's Edition**, look for **Differentiated Instruction** boxes. Many of these boxes contain activities specifically designed for English learners.

Standardized Test Practice

Chapter Resource File 8

HOLT Science Spectrum

Physical Science

Solutions

Chapter Test

Waves

Chapter Test

Forces

The **Chapter Resource Files** provide differentiated worksheets, including **Chapter Tests,** and **Standardized Test Practice with Guided Reading Development.**

Teaching the Language of Science

▶ Why Is It Important?

English language learners need to experience the excitement of science, the journey from inquiry to discovery, while at the same time learning how to communicate their discoveries in the language of science. They need to learn the language of science as they simultaneously develop their conceptual understanding of science. Integrating language and science instruction is challenging. However, if students do not learn the language of science, they cannot access their textbooks, pass science exams, or master science standards.

▶ Strategies and Resources

The best place to teach the language of science is in the context of science lessons. It is easy to become so focused on science standards that you only teach science concepts and you overlook teaching the language of science. It should not be difficult, however, to add a focus on language in science teaching.

Word instruction is taught throughout the program—in the Student Edition, the Teacher's Edition, and the Interactive Reader. In addition, you can give students tips for planning the overall organizational structure of essay answers and for using vocabulary words accurately. Ideas for language objectives are found throughout the Teacher's Edition.

In the **Student Edition,** scientific vocabulary terms are highlighted and defined within the text, and definitions are repeated in the margins.

The Reading Toolbox in the **Student Edition** contains segments that will help students understand the meaning of scientific vocabulary. For example, some activities define prefixes, roots, and suffixes and other activities explain word origins.

Appendix A in the Student Edition contains a table of word parts that make up common scientific terms. **Appendix A** also includes a table that shows everyday words that are used in science.

Reading Toolbox activities are included throughout the chapter in both the **Student** and the **Teacher's Editions** to reinforce the language tools learned at the beginning of the chapter.

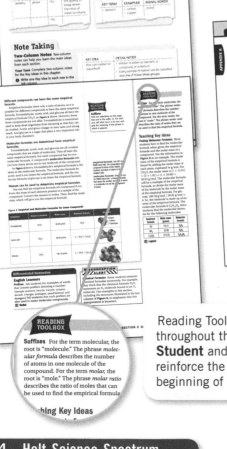

Modeling Academic English

▶ Why Is It Important?

Students who are surrounded by other English language learners may not get much exposure to academic English. In class, students tend to use their first languages or an informal variety of English. It is challenging to provide students with adequate models of academic English. To acquire academic English, students need to have many exposures to the features of academic English and multiple opportunities to use this variety of English.

▶ Strategies and Resources

You can increase the students' exposure to academic English by making additional reading assignments and asking students to watch educational television shows and movies and to listen to specific news broadcasts. In class, you can establish clear rules concerning language use in activities designed to build English language proficiency. Modeling and scaffolding language use before partner and small-group activities increases students' opportunities to convey their ideas about science in academic English. Giving students reading passages to refer to when they complete group activities encourages the students to incorporate words and expressions from the passages into their own speech.

Academic terms are taught throughout the **Student Edition. The Reading Toolbox** teaches students types of academic language that they will encounter in the chapter. Students are then assessed on their understanding of vocabulary in the **Chapter Review.**

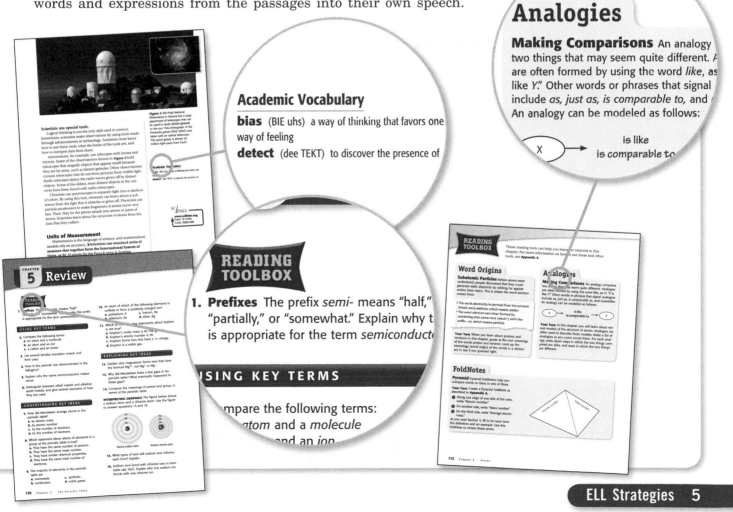

Academic Vocabulary

bias (BIE uhs) a way of thinking that favors one way of feeling

detect (dee TEKT) to discover the presence of

Analogies

Making Comparisons An analogy two things that may seem quite different. are often formed by using the word *like,* as like *Y.*" Other words or phrases that signal include *as, just as, is comparable to,* and An analogy can be modeled as follows:

X → is like / is comparable to

READING TOOLBOX

1. **Prefixes** The prefix *semi-* means "half," "partially," or "somewhat." Explain why t is appropriate for the term *semiconduct*

▶ Why Is It Important?

Science reading is challenging because it contains both complex language and abstract concepts that English language learners can not understand. All students must access the reading to understand their science lessons, develop their knowledge of science, and reach challenging standrds..

▶ Strategies and Resources

You can increase your students' reading comprehension while helping students reach science standards. Take advantage of Holt's interactive materials, use strategies and activities designed to teach students the language of the textbook, give students opportunities to practice using this language, employ visual aides, allow students time to read, give students opportunities to summarize their reading, assess student comprehension frequently, and give students opportunities to clarify their understanding.

Each chapter begins with a **Reading Toolbox** that provides a reading comprehension strategy for students to use before they begin reading the chapter. Comprehension questions are designed to tap students' understanding of the readings. Critical-thinking questions are provided in both the Student Edition and the Interactive Textbook. These questions are designed to encourage the development of critical analysis and abstract thought.

The **Reading Toolbox** at the beginning of each chapter contains instructions on how to use Graphic Organizers and Note Taking strategies. The **Reading Toolbox** activities in the chapter show students how to use these tools to improve their comprehension and retention of the material. Each section in the chapter also contains multiple **Reading Checks** so that students can check their understanding of the material as they read.

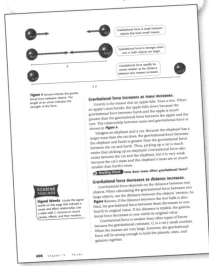

a new way to read

Live Ink Reading Help, a scientifically-researched tool for improving students' reading comprehension, is available on the **Interactive Online Edition** of *Holt Science Spectrum.*

The **Interactive Reader** includes frequent **Reading Checks** and **Critical Thinking** questions that test whether students comprehend what they are reading.

The **Chapter Resource Files** contain **Concept Review** and **Cross-Disciplinary** worksheets that provide students with additional opportunities to improve their reading comprehension.

Challenge: Developing Oral Expression

▶ Why Is It Important?

Engaging English language learners in oral activities is challenging. English language learners may lack the confidence to speak in front of their peers, or they may fear ridicule or evaluation of their speech. They may not be accustomed to expressing opinions or arguing a point in English. Participating in well-structured oral activities helps motivate students, clarify their understanding of science, and teach concepts and language simultaneously.

▶ Strategies and Resources

You can maximize English language learner participation in oral activities by arranging frequent pair and group activities focused on instructional tasks, by modeling oral responses, by clarifying and assigning tasks that require student participation, and by teaching students ways to ask questions, request clarification, and participate in specific science activities.

One of the best activities for developing oral language is summarizing. Ask students to summarize specific passages in the materials in pairs. First, give them time to study a particular passage. Then, have them summarize the same passage more than one time. This strategy builds their confidence and fluency. You will find that they use the words in the passage in their speech, which helps them understand the words and acquire them.

You can also structure short one- to five-minute oral presentations in ways that guarantee the development of oral expression. Give students plenty of time to prepare. Require them to base their presentations on reading material, and give them lists of useful expressions and sentence starters as well as tips for making their presentations. Tell them to use visual aides. Have them practice their oral presentations aloud with a partner before they make their presentations to the class. To ensure that all students are engaged, you may want to ask all presenters to prepare one or two engaging questions that they can ask their classmates when they have finished their presentations.

Talk About It elements in the **Interactive Reader** provide crucial opportunities for English language learners to practice section vocabulary.

Talk About It

Brainstorm In a small group, think of some examples of situations in which you shouldn't use the height of an object above the ground to calculate its gravitational potential energy.

...in pairs. List the symbol... ...d mass numbers of 20 elements on the chalkboard. Have students create a table that lists the number of protons (atomic number) and number of neutrons (mass number–atomic number) for all of the elements listed. Then, ask them to summarize the information in their table. (For light elements, the number of protons and the number of neutrons are approximately equal. For heavier elements, the number of neutrons increases faster than the number of protons in an atom.) **LS Verbal**

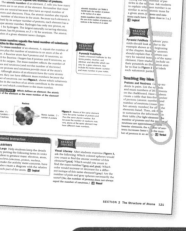

The **Teacher's Edition** contains a variety of activities labeled **LS Verbal** that engage students in making oral presentations.

▶ Why Is It Important?

Students need to receive feedback on their oral assignments, even when they are still acquiring proficiency in English. To continue to acquire English, they need to know both their strengths and weaknesses in English.

▶ Strategies and Resources

Detailed feedback on your students' ability to express themselves orally is not required. However, it is a good idea to give students two grades on their oral work, one on their content and the other on their general use of English. You can use the rubrics shown below, which can be found in the Program Introduction Resource File booklet. You may also want to use the oral rubric on the following page when grading oral presentations.

The **Program Introduction Resource File** booklet contains a wide variety of rubrics and assessments, including the rubrics shown here. Use the teacher evaluation rubrics to evaluate how the oral presentation skills of English language learners improve throughout the school year. You can also use the self-evaluation and peer evaluation rubrics to help students learn how to evaluate their own oral presentation skills.

Name: _____

Oral Presentation Grading Criteria

Aspects of good oral presentations	Scoring:	Your score:

❶ Introduction *(30 points possible)*

- Introduce topic clearly
- Tell what you will discuss
- State main idea
- Provide topic background

27–30 Exceptional
24–26 Very Good
21–23 Average
0–20 Needs Improvement

_____

❷ Body *(25 points possible)*

- Make main points clear
- Fully support main ideas
- Use well-planned organization
- Use effective connections
- Support main idea well

23–25 Exceptional
20–22 Very Good
17–19 Average
0–16 Needs Improvement

_____

❸ Conclusion *(10 points possible)*

- Reinforce central idea

9–10 Exceptional
8 Very Good
7 Average
0–6 Needs Improvement

_____

❹ Delivery *(20 points possible)*

- Maintain eye contact
- Avoid distracting mannerisms
- Articulate words clearly
- Use grammatical English
- Speak at a natural pace

18–20 Exceptional
16–17 Very Good
14–15 Average
0–13 Needs Improvement

_____

❺ Visual Aids *(15 points possible)*

- Contribute to presentation
- Neat
- Accurate
- Clear

14–15 Exceptional
12–13 Very Good
10–11 Average
0–9 Needs Improvement

_____

Total Points _____

Presentation Grade _____

The **Bellringer Transparency** to be used at the beginning of each section helps students to focus on concepts and scientific language. **Concept Mapping Transparencies** give students practice organizing ideas and concepts.

Graphic Organizers and **Note Taking** help students learn and practice note-taking techniques.

Why Is It Important?

English language learners often lack the listening skills required to understand their teacher's oral explanations and presentations. The inability to understand their teacher and classmates undermines their success in science classes. It can even prevent English language learners from learning key concepts, completing assignments correctly, and developing their English.

Strategies and Resources

You can help students improve their listening skills by providing them with Graphic Organizers during oral presentations of materials. Ask the students to complete the Graphic Organizers as you present an explanation of a science concept orally. For samples of Graphic Organizers, draw upon those provided in Appendix A.

Note taking is an excellent way for students to improve their listening skills. Ask students to listen for specific words, explanations, descriptions, examples, and definitions. Encourage them to outline the key points. Various methods for note taking are described in Appendix A.

To improve the listening comprehension of your beginning students, speak slowly and clearly; use short, simple sentences; employ visuals; demonstrate key ideas; and use gestures. It is also helpful to paraphrase and repeat ideas in different ways.

Why It Matters topics provide topics of discussion that engage and attract students to the science content of the section.

Reading Toolbox Visual Literacy items allow students to listen to explanations and to participate in discussions about interesting art and photos within each chapter.

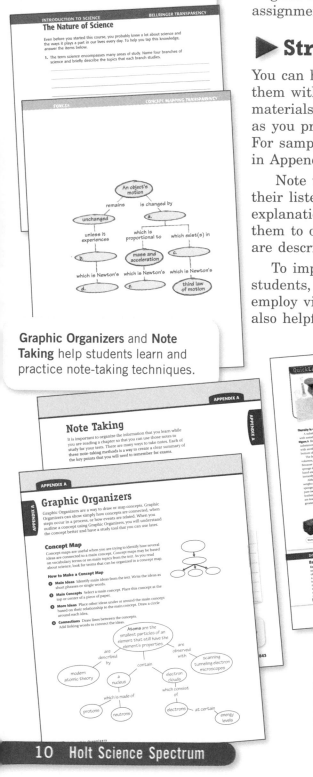

Challenge: Evaluating Listening Ability

▶ Why Is It Important?

When trying to evaluate your students' listening skills, simply asking the class if anyone had difficulty understanding you will not work. Students are often reluctant to reveal their language comprehension problems to classmates. However, it is important to determine how well your students understand the science concepts discussed in class. Evaluating the listening ability of your students will help the class proceed smoothly by enabling you to review ideas that your students do not understand, to build upon these ideas, and to challenge your students to examine their understanding of the world.

The **Guided Reading Audio Program** can help struggling readers improve their listening skills while helping them understand the content in the textbookl

▶ Strategies and Resources

You can evaluate students' ability to understand your oral presentations by giving students a quick quiz or activity after you have discussed a concept. *Holt Science Spectrum* provides many such quizzes and activities, including the Section Reviews, Chapter Reviews, quizzes, and tests, as well as Standardized Test Preparation worksheets. You can also give students a true/false quiz concerning details covered in lectures or class discussions. Simply write a list of 15 statements (both true and false) and ask students to respond to the statements. To verify their understanding, you can ask students to write brief summaries of the information that you have just given to them or you can ask students to complete Graphic Organizers.

Section Quizzes also assess the degree to which your students effectively listened during class discussions.

Concept Reviews provide quick ways to evaluate the listening comprehension of concept-specific discussions.

Developing Students' Written Expression

▶ Why Is It Important?

Students learn and display their knowledge of science through writing. Throughout the course of a science class, students will need to write reports, notes, summaries, answers to questions, and even essay-like passages. Yet, of all the language skills—reading, writing, speaking, and listening—writing is the most difficult for English language learners.

▶ Strategies and Resources

You can help students develop writing skills by implementing some or all of the following strategies. Provide instructions that are closely aligned to the specific writing assignment. Before giving any writing assignment, explain the assignment clearly, give the students some tips on getting started and organizing their essays, discuss some of the language the students might need to complete the assignment, and share sample writing assignments and grading criteria. Suggestions for providing feedback on student writing are provided below.

Students can practice their writing skills through a variety of activities. Ask students to answer questions with single phrases, sentences, or paragraphs; to list the main ideas of passages; to summarize sections; to react to your lessons; to report on investigations; and to complete essays. Give short, ten- to twenty-minute writing assignments in class that the students can later revise and edit at home. Assign short writing assignments for homework. Teach students to write longer assignments in stages. Encourage students to take notes in their own words as they read.

Throughout the *Holt Science Spectrum* program, there are numerous writing assignments. For instance, students are asked to respond to critical-thinking questions, write reports, complete Graphic Organizers, and take notes.

The Critical Thinking questions in the **Student Edition** give students an opportunity to practice writing about the major concepts in the chapter.

Analyzing Data A jar contains 30 mL of glycerin (mass = 37.8 g) and 60 mL of corn syrup (mass = 82.8 g). Which liquid is the top layer? Explain your answer.

21. **Applying Concepts** A light green powder is heated in a test tube. A gas is given off while the solid becomes black. What type of change is occurring? Explain your reasoning.

22. **Making Inferences** Suppose you are planning a journey to the center of Earth in a self-propelled tunneling machine. List properties of the special materials that would be needed to build the machine, and explain why each property would be important.

WRITING IN SCIENCE
1. Franklin is credited with much work in addition to his groundbreaking electricity experiments. Prepare a presentation in the form of a skit, story, or computer program about his work on fire departments, public libraries, or post offices.

Many of the questions in the **Why It Matters** articles, including the **Writing in Science** questions, provide opportunities for students to express their knowledge of science.

Improving Math Literacy

▶ Why Is It Important?

Learning to read physical science has an added challenge because students must also master the language used in math. Understanding how to read math problems is essential for success in science. Some English learners may excel in math, but cannot show their strengths if they do not understand what a problem is asking. English learners who struggle with math concepts will have difficulty improving their math skills if they do not know how to begin approaching a problem. Students who do not learn to apply basic math skills will not be successful in higher-level science classes. It is important to identify whether students are struggling with math problems because they do not understand the wording of a problem or because they need to develop their math skills.

▶ Strategies and Resources

You can improve students' math literacy by working through word problems stepwise. The Math Skills segments within the chapters help students identify the important information in word problems and break down the problems into manageable steps. Students can see how to work through the type of math problems that they will need to solve to master a chapter's content. The Math Skills on the Science Skills pages in some chapters show students how to apply common math skills to science problems.

To determine if students are stuggling with the basic math skills required to solve science problems, have them solve simple math expressions and algebraic equations, instead of word problems. Appendix B provides tables of common rules used to solve equations, as well as practice solving basic math expressions and algebraic equations.

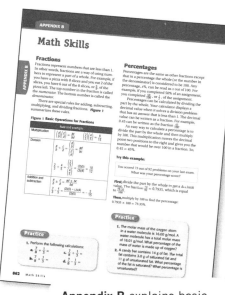

Appendix B explains basic math skills to students and provides opportunities to practice these skills.

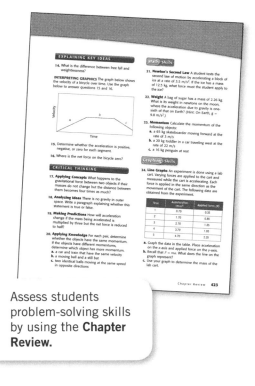

Assess students problem-solving skills by using the **Chapter Review.**

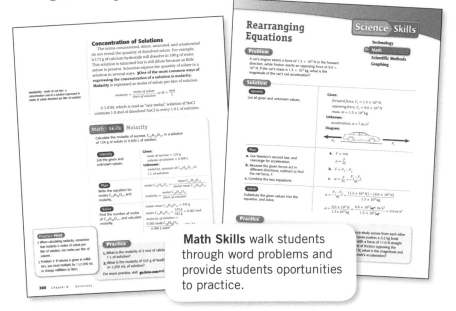

Math Skills walk students through word problems and provide students oportunities to practice.

Providing Effective Feedback

▶ Why Is It Important?

It can be time consuming and difficult to provide students with effective feedback on their writing assignments. Without this feedback, however, their English language development may cease, and they may never attain academic literacy.

If effective feedback is not provided in earlier grades, a serious problem can arise when English language learners reach secondary school and are required to use academic English accurately to do well on tests and to use academic English in their writing. Their knowledge may be so far below that which is required to complete writing assignments that they are not able to catch up to their peers without intensive instruction.

▶ Strategies and Resources

In the earlier grades, students may not have received comments on language problems, or they may have received only global remarks, such as "Keep working on your grammar." To help students acquire academic English and to correct patterns of error that may persist in their writing, teachers need to provide structured help. However, providing the correct forms when the students are capable of making their own corrections is not the answer. Below is a list of suggestions initially developed by Susan Earle-Carlin for university students and then revised by Tustin and Calexico Middle School science teachers.

- Try to predict potential errors and to supply correct forms before students complete a writing assignment. For example, if the students are writing about erosion, remind them that *erosion* is generally a non-count noun (and thus cannot be made plural by adding an -s ending) and that the word has a related verb form *erode*. Also, tell them that the correct phrase is "Soil <u>erodes</u>." (It does not "fall apart.") Providing information about the language that students need to know to complete writing assignments should reduce errors.

- Plan to give students time to proofread their lab reports and other writing assignments in class before handing them in. This process models good behavior and encourages students to ask questions and use their dictionaries.

- If you use correction symbols, use them consistently. Teach the students what the symbols mean. Have students work in groups to practice editing a sample paragraph marked with correction symbols; then give students time to ask questions when you return their work. Spanish-speaking students in the advanced stages of English language development will need help with articles, auxiliary verbs (particularly forms of *do* and the modals, *must, may, might, should, shall, could* and *can*), prepositions (particularly following verbs and adjectives), and sentence structure.

- Mark errors in ways that demand more of the student's attention in each writing assignment as the school year progresses. Here are some suggestions: First, circle or underline errors and write the correction symbol above them. Next, highlight the errors without supplying the symbols. Then, write only the symbols in the margin of any line with the errors. Finally, put only a check in the margin to indicate that there are errors of some sort.

- Supply quick margin notes for underlined problems, or refer the students to a learner's dictionary.

- Develop a system to provide set phrases, either in the margin or in footnotes or endnotes, or refer the student to a learner's dictionary. If a student has written a non-idiomatic phrase, it is not always enough to underline or highlight it. English language learners have often misheard set expressions and cannot figure out what is wrong. For example, they may use the expression, "on another hand" instead of "on the other hand."

- Provide motivational feedback. Remember that a key goal of helping English language learners is to keep them interested in writing about science.

Examples of written feedback appear below.

> **I like your description of erosion.**
>
> **You write interesting introductions. I like the way you put your thesis statement in the last sentence of the first paragraph.**
>
> **I like the way you shift verb tenses here.**
>
> **Only your subject-verb agreement stands in your way to tremendous success as a writer.**

This method of making the student more responsible for recognizing errors can be spread out over the school year (ranging from symbols over errors in early assignments to checks in the margin at the end of the school year). For intermediate and advanced English learners who are producing unacceptable work, it is useful to continue to provide clear feedback by using the correction symbols over each error.

Below is a list of correction symbols that you may want to use. Notice that the first column contains the symbol, the second column explains the meaning of the symbol, the third column contains an example, and the fourth column contains the corrected sentence.

Correction Symbols

Symbol	Meaning	Example	Corrected Example
agr	agreement	Each biologist has their *own* problems.	Each biologist has his or her own problems.
cap	capitalize	*lake erie* has a volume of about 483 trillion liters of water.	Lake Erie has a volume of about 483 trillion liters of water.
cs	comma splice	The earthquake happened quickly, the people there ran away.	The earthquake happened quickly, and the people there ran away.
frag	fragment	If he were here.	If he were here, we could complete the project.
id	idioms / set expressions	He was involved *on* a study of bacteria.	He was involved in a study of bacteria.
num	number	The computer helped them to do their *works*.	The computer helped them to do their work.
p	punctuation	They had stems and, branches.	They had stems and branches.
red	redundancy	The *concluding* result was not what we expected.	The result was not what we expected.
ref	unclear pronoun reference	The scientist undertook the study. They were wrong.	The scientist undertook the study. The researchers were wrong.
ro	run-on	No one knows what is happening to the ozone it is hard to understand what is happening.	No one knows what is happening to the ozone, and it is hard to understand what is happening.
sp	spelling	Understanding science is *important*.	Understanding science is important.
s-v	subject-verb agreement	Everybody *have* reasons.	Everybody has reasons.
t	tense	New studies arise last year.	New studies arose last year.
vb	verb form	The scientist liked use powerful lens.	The scientist liked using a powerful lens.
wf	word form	We became *independence* thinkers.	They dabbed powders on the paper.
ww	wrong word	He was *very* tired that he left.	He was so tired that he left.
^	insert	They dabbed powders ^ paper.	They dabbed powders on the paper.
℮	delete	They also studied insects too.	They also studied insects.
//	parallelism	Studying and *replicate* is part of her job.	Studying and replicating is part of her job.
#	add a space	It is *infront* of the tree.	It is in front of the tree.
~	rephrase	They didn't get a clue about the new study.	They did not understand the new study.
??	not understandable	Understanding geologic processes helps scientists the features of Earth, but of other planets.	Understanding geologic processes helps scientists explain the features of Earth.

Here is an example of how one teacher marked her student's writing:

ww
<u>Volcano</u> eruptions are powerful and destructive

num.
<u>force.</u> Imagine hearing a volcano erupt

thousands of miles away.

volcano = noun
volcanic = adj.

v + ing
Imagine <u>look</u> through binoculars and <u>see</u> the

v + ing

top of a mountain collapse.

The word *imagine* is followed by a verb and *-ing*.

cap
Imagine discovering an ancient <u>r</u>oman city

ww
buried in <u>volcano</u> ash.

Don't forget to capitalize proper nouns.
You have some great ideas. Only your grammar stands in
your way to tremendous success as a writer!

Teaching Reading to
ENGLISH LANGUAGE LEARNERS

in the Science Classroom

by Dr. Robin Scarcella

English Language Learners

English Language Learners (ELLs) are students who are learning English and who speak a language or dialect other than English. They require additional English language support materials to access their reading and participate in their science classes. Between 1990 and 2005, the number of public school students not fully proficient in English increased by more than 100 percent to reach more than 4.6 million. These students come from many different language, educational, and socio-economic backgrounds. They speak over 460 languages, though Spanish is the native language of about 79 percent of the ELL population in this country. Some ELL students have strong first-language and literacy skills, though most do not. Some are recent arrivals to the United States, but many are long-term residents of the United States who speak a variety of *learner English.*

Use Caution Ahead

Just because students are able to communicate orally does not mean that they have acquired enough English to succeed in your science class.

Identifying ELL Students

One of the first things you can do to help English Language Learners is to identify their English language proficiency levels with diagnostic assessment. On the first day of class, give students an informal writing assessment, asking them to write a paragraph using complete sentences to describe their previous science classes. Have them give you specific details about their science coursework, including the topics they studied, the approximate years they studied these topics, their previous schools, and the locations of these schools. This informal writing assessment will provide important information about the students' previous education, both in the United States and other countries, and will give you a sense of each student's level of English mastery.

To assess students' reading skills, ask them to read a short passage from the textbook and then answer comprehension questions. Although this assessment does not cover all of the language skills (reading, writing, speaking, and listening), these informal assessments will help you categorize the students into three main proficiency levels: beginning, intermediate, and advanced. Chart 1 provides suggestions for analyzing the students' work.

Chart 1: Determining the English Proficiency Levels of ELL Students

English Proficiency Level	Diagnostic Writing Assessment Results	Diagnostic Reading Assessment Results
Beginning	unable to complete assignment or able to write only short or incomplete sentences with highly limited vocabulary and many errors; writing may be incomprehensible, unable to use past tense verbs, such as *learned, studied, took,* and *went*	unable to read science material and to answer questions pertaining to the material
Intermediate	able to complete assignment without many language errors; uses past tense verbs incorrectly or inconsistently	able to answer some questions, though answers may not be completely accurate
Advanced	able to complete the assignment with some language errors; able to use many words correctly; uses the past tense accurately	able to answer most of the questions, though some answers may be incorrect

Use Caution Ahead Helping students understand difficult language is far more challenging than simplifying the language and concepts for students.

Beginning English Language Learners

Learners who are just beginning to acquire English have difficulty writing even the most basic sentences in English. They write short sentences, sometimes just a word or two in length, and make many errors. Sometimes they are unable to read any of their science material in English. They have trouble understanding others without enormous effort. They have difficulty communicating in spoken English; their speech lacks fluency and often contains many pauses.

While these students should be attending special English language-support courses or newcomer programs, you may find them in your science classrooms. Your primary role is to improve their understanding and use of Basic English, the English used in most everyday situations. In addition, you can teach them science terms and expressions and can help them understand the foundational concepts of their texts.

Intermediate English Language Learners

Most of the English Language Learners enrolled in middle school and high school have an intermediate level of English proficiency. You can identify these students by their ability to communicate virtually anything they want in everyday situations—both in speech and in writing—and by their inability to communicate their ideas in academic situations, such as scientific inquiry. Their writing, which is composed mainly of simple sentences, is often riddled with errors. They can use everyday words but have difficulty using scientific terms accurately. Learners with an intermediate level of English proficiency are able to read basic materials (the types that younger students might read), but they find reading science texts enormously challenging. They read these texts inaccurately, often glossing over difficult terms, misunderstanding sentence structures, and failing to notice logical connections between words.

Without sufficient instruction, these learners often give up trying to access scientific text. They may rely on their teachers' explanations of readings, pre-reading activities, titles, headings, and illustrations to guess the content of the readings. These learners never acquire enough English to graduate from high school or go on to college.

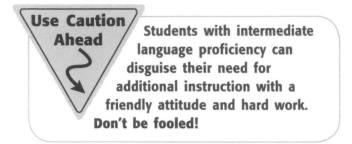

Use Caution Ahead Students with intermediate language proficiency can disguise their need for additional instruction with a friendly attitude and hard work. Don't be fooled!

Advanced English Language Learners

Learners with an advanced level of English are able to communicate mostly everything they want to say in both informal and academic situations. They may even excel on standardized tests of content knowledge. However, when they speak and write, they continue to make some grammatical and vocabulary mistakes. Without instructional support, their grammar and vocabulary do not improve. Although they are able to read science texts with effort, they have difficulty understanding subtle meanings.

Teaching Students to Read Science Texts

Teaching students to read their science materials will give them a fighting chance of reaching challenging standards and passing high-stakes tests. This kind of reading is developed through instruction. You can draw students' attention to the ways in which language is used in their textbooks, giving them strategies for learning the language and providing them with a lot of practice.

The National Reading Panel recommends that students receive instruction in the areas of phonemic awareness, phonics, reading comprehension, and fluency. Recent second-language research supports this recommendation.

Regularly Assess Progress!

In considering your students' proficiency levels, keep in mind that ELL students are capable of making remarkable progress in acquiring English. Their language proficiencies can change quickly. As a result, you need to assess their ability to handle scientific reading periodically throughout the year.

Phonemic Awareness and Phonics

Phonemic awareness consists of identifying and manipulating the sounds of a word, while *phonics* consists of associating letters with sounds. Both strategies aid reading and spelling. Most English Language Learners already have these skills, but some may not. Even adolescents who have lived in the United States all their lives sometimes lack these skills. While you are not responsible for teaching these skills, you are responsible for identifying learners who lack phonemic awareness and phonics skills and for referring them to your school reading specialist. The specialist can offer students academic support and can provide you with suggestions to help them.

You can identify these students by their inability to read even basic passages. These learners struggle to make sense of the letters on the page. They cannot sound out words accurately. They cannot write.

Reading Comprehension

Research shows that students develop reading comprehension through a combination of reading comprehension strategy instruction and language instruction. Reading comprehension strategies are conscious plans or sets of specific steps that enable students to make sense of concepts and remember them. These strategies help students identify and "fix" their reading comprehension difficulties. You can teach these strategies explicitly, explaining why the strategies are useful and when and how to use them. After you explain the strategies, take time to demonstrate the ways you use them and talk about their use as you model them. When appropriate, provide students with useful lists of words and expressions they might need to use the strategy. Then give students guided and independent practice.

Before giving a reading assignment, you must assess students' knowledge to discover what students already know about concepts and what they need to know. Also provide instruction on foundational concepts and teach them strategies to learn these concepts. One group of strategies, *Before Reading Strategies*, helps build the students' foundational knowledge of the concepts in the reading material before the students have read it.

During Reading Strategies include strategies that students can use when they are reading, such as asking questions about their understanding of the text, adjusting their reading speed to fit the difficulty of the text, reading the text in reasonable sections, and identifying and fixing their comprehension problems. Remind students that science text is difficult and that they may have to read and outline sections many times to understand the material. *After Reading Strategies* include asking students to check their understanding of what they have read.

BEFORE READING STRATEGIES

Students can do the following:

- ✓ *Clarify the purpose,* or explain their reason for reading.
- ✓ *Preview* the text by looking at headings and illustrations to anticipate the content.
- ✓ *Ask questions* about what they already know about the selection content.
- ✓ *Brainstorm* with you and the rest of the class to discuss what they anticipate the reading will be about and build their knowledge of the subject. You can write useful words and expressions that come up on the board.

DURING READING STRATEGIES

Students can do the following:

- ✓ *Look for answers* to questions.
- ✓ *Identify difficulties,* including noting the passages, sentences, and words that are confusing.
- ✓ *Restate* difficult passages in their own words.
- ✓ *Review and look ahead* to figure out the specific parts of the text they don't understand.
- ✓ *Use graphic organizers* such as maps, diagrams, graphs, charts, flowcharts, and cluster charts to identify and understand critical science concepts and their relationships to other concepts.

✓ *Summarize.* Summarizing requires students to identify key concepts in their reading and connect these concepts to others, to eliminate superfluous information, and to express information concisely in their own words. Whether completed orally with a partner or in writing, summarizing helps students remember what they read while they acquire English.

✓ *Discuss.* Guide students in discussions of the text, discussing the reading, clarifying and confirming student knowledge of difficult concepts, or focusing on one or two strategies students have used to comprehend difficult concepts. Provide ELL students with useful lists of words and expressions they can use in these discussions.

✓ *Generate and answer questions.* Check and confirm students' knowledge of difficult concepts by having students ask and answer questions about the material, using sample questions and a set of expressions that you give them. Students could complete this activity in pairs or small groups.

Language Instruction

Reading comprehension improves significantly with language instruction. The language of science textbooks is often so unfamiliar to students that even native English speakers sometimes feel they are learning a foreign language. You can imagine how difficult science passages can be for English Language Learners who are still in the process of grappling with the English of everyday, ordinary situations.

Types of Science Writing

Teaching the various types of writing used in science textbooks demystifies science passages. Competent readers understand these writing types, their purposes, organization, and key language features. One important type of writing is **factual writing.** To understand this type of writing, competent readers know

- language for introducing major points (e.g., *one important, another important, yet another, in addition,* and *a further consideration*)
- language that indicates examples, illustrations, facts, and statistics to support the points made
- logical connectors (such as *first, second,* and *next*) that provide smooth transitions between words and sentences

Another important type of writing is **assessment writing.** Introducing ELL students to the predictable language patterns found in tests will do much to build their confidence and reading proficiency. To get students ready for tests, regularly discuss typical test items that contain short reading passages. Analyze the reading in the passage for the students, teaching them ways to locate the main ideas. Also unpack the language to help students understand the meaning of the reading items. Model strategies that students can use to answer the questions.

Vocabulary

Reading comprehension improves greatly when students expand their vocabularies. To succeed in science, students must become familiar with several types of vocabulary:

Common vocabulary with specialized meaning. Many words from everyday vocabulary have specialized meanings in science. Consider the words *fault, reflection, power, force, active,* and *plate.* Even common words can take on a very precise and possibly unfamiliar meaning in science classrooms; for example, the use of the preposition *by* to mean "according to" in the sentence "I want you to sort **by** color."

Academic vocabulary. Academic vocabulary includes words that are common in academic texts of arts, science, law, and commerce.[1] These words are different from the technical or content terms used in science (e.g., *cell, nucleus,* and *molecule*). They are words used across all academic disciplines (e.g., *accommodate, inhibit,* and *deviate*). Coxhead provides a list of 570 academic word families identified from an academic corpus of 3,500,000 words.[2] Academic words are difficult in part because they appear relatively infrequently, they often occur within contexts that are less familiar to students, and they are cognitively complex and abstract.

Content vocabulary (sometimes called technical words). Content words such as *germinate, piston,* and *stamen* are discipline-specific words. In science, these words can entail an understanding of other content words. To understand the term *ecosystem,* for example, readers must understand other important concepts. Such concepts include *predator, prey,* and *decomposers.* Knowing the technical meaning of science words sometimes entails understanding whole taxonomies.

Use Caution Ahead

Guessing a definition from the way the target word is used in a sentence helps students figure out the meaning of the word only when they know at least 90 percent of the words that surround the key word.

Strategies for Teaching Vocabulary

Present vocabulary orally. Students repeat each word after you two or three times. If the word is more than one syllable, sound out each syllable, pointing out the syllable(s) the students should stress. This exercise gives students the opportunity to hear and practice the word and its stress pattern.

Examine the spelling of the word. Point out interesting letter combinations and spelling patterns.

Explain spelling etymology. Atmosphere has a *ph* in it. The *ph* sounds like [f]. When *ph* sounds like [f], it comes from Greek.

Present the word orally in a sentence. Read the sentence from the text, and have your students repeat the sentence orally. This practice allows students to hear and use new words in sentences.

Define the word. Explain the meaning of a new word in one or more of these ways; presenting the definition of the word to the students or eliciting the definition of the word from the students:

1. *reading* the definition from the textbook;
2. *paraphrasing* the definition (restating the definition in other words);
3. *showing* pictures that illustrate the meaning of the word;
4. *analyzing* word parts;
5. *prompting* students to guess the meaning of the word from the context in which the word occurs;
6. *explaining* the word through pantomime (for example, to explain words such as *surround* when explaining the terms *atmosphere* and *outermost layers*).

Explain other word meanings. For example, you might say, "In science, the word *atmosphere* refers to the air that surrounds the earth. However, atmosphere can also refer to an environment or describe a setting. For example, we can describe the atmosphere at a baseball game as casual, a wedding as joyful, or a slumber party as noisy."

---

[1] Nation, 2001, p. 15. (Complete references can be found in *Section III: Continuing the Discussion,* p. 171.)

[2] Coxhead, 2000.

Vocabulary Activities & Games

The "Where Would You Find X?" Game

Humidity?
Precipitation?
A current?

Students answer the questions orally, and you discuss the answers with them.

Correcting Mistakes

Ask students to correct mistakes in a series of sentences. For instance, write on the board a series of sentences that contain mistakes. One example is: ***Mountains can be found in the stratosphere.***

Paraphrasing

Ask students to paraphrase the definition of the words. Students can write their own definitions.

A Simple Recall Game

Provide students with a definition, and students supply the word. In *cloze* exercises, you delete words from a text that the students have already read. You delete every 6th word or a limited number of target words you want students to know. When making the deletions, keep at least the first and last sentences intact, giving the students the opportunity to figure out the context of the reading. Here is a sample oral cloze activity: ***A wind that affects a small area is a _____.*** (Snap your fingers, and students reply.)

Word Learning Strategies

To help students remember targeted words, employ a variety of strategies, including visualization, vocabulary note cards, and word grids (including the word, the definition in the students' own words, the students' original sentence, and an illustration or picture when possible), and Question Grids, in which students answer questions such as ***What is the word? What does it mean? What is it like?*** or ***What are some examples of the word? How is the word used in science?***

True/False Activities

Students can complete the activities in writing, marking ***T*** for true and ***F*** for false, or can complete the activities orally, calling out ***true*** if the statement is true and ***false*** if the statement is false.

Sentence Dictation

Give students sentence dictation activities. These types of activities are rarely given in science classes, but they are a big help to ELL students who are developing word knowledge.

Graphic Organizers

Discuss relevant graphic organizers with the class, having the students fill in missing information to clarify word meanings.

These activities and games are examples of strategies you can use to help students build their vocabulary.

Posters and Word Walls

Post new words and their definitions and sentences. Students discuss the words briefly whenever they come up. Or hang mobiles with cards that contain the essential science words and sample sentences.

Word Games and Puzzles

A number of word card games can be played in pairs, in groups, and as a class. For instance, you can engage students in sorting activities in which students categorize words, sorting and grouping them by similarities and differences. In a closed sort, you provide the categories. In an open sort, the students generate the categories. You can also have students work in pairs, giving each student in the pair a different set of words on word cards and telling students to take turns guessing each other's words. In this activity, students ask whatever questions they want.

The "Who Am I? What Am I?" Game

You ask questions, and the students respond out loud chorally. For example:

What am I? I blanket the Earth. I consist of gaseous material.
What am I? I am closest to the Earth's surface.
What am I? I push on everything all the time.
What am I? I am caused by the rising and sinking of air.
What am I? I am the weight of the atmosphere caused by gravity. (Gravity is . . .)
What am I? I am a horizontal movement of air that affects a small area.
What am I? I am a horizontal movement of air that affects large parts of the Earth.

Making Personal Connections

Ask students to make personal connections to new words. For example, ask, *Who in this class has seen a snake slough?* If students write their responses, take the time to provide them with supportive corrective feedback.

Matching Activities

Ask students to match words to their appropriate definitions.

Yes/No–Either/Or Questions

You ask students a series of yes/no and either/or questions about the word. For example, ask, *Is the Earth surrounded by a blanket of air or dirt? Is air pressure the weight of the atmosphere due to gravity or due to . . .?* These activities can be oral with a choral or written response.

Grammar

Reading science texts competently requires a mastery of grammatical features that express relationships in direction, order, quantity, shape, size, and space. Not knowing these features can make science text impossible to understand. Even if you yourself do not know much grammar, you still have a sense of the ways grammar works to convey meaning. Simply writing sentences on the board and calling attention to their patterns and punctuation will help students learn their grammatical features. The following aspects of grammar are especially challenging to English Language Learners as they learn science: sentence structure, verb tense, passive structures, prepositions, adjectives and adverbial forms, and pronouns.

Sentence Structure. Each sentence has a specific structure that conveys meaning. Complex sentences often represent abstract concepts. Before making a reading assignment, help students understand a complex sentence structure that might confuse them when they encounter it in their reading. Write sentences from the textbook that contain these structures on the board. Prompt students to break down complex sentences into smaller sentences that mean the same. For example, consider how the following sentence can be broken down.

> ***When an electron is added to or taken away from an atom, the atom becomes a charged particle called an*** ion.

✔ An electron is added or taken away (subtracted).

✔ At this time, the atom becomes charged.

✔ Then, the atom is called an *ion*.

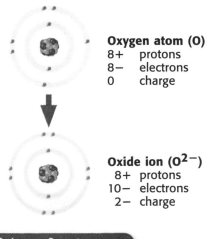

Oxygen atom (O)
8+ protons
8− electrons
0 charge

Oxide ion (O^{2-})
 8+ protons
10− electrons
 2− charge

Another example of a complex sentence is ***Because the air temperature and pressure vary so much over the Earth, the air in the troposphere is constantly moving***. Notice that the two clauses of this sentence are connected by the word *because* at the beginning of the sentence.

Common words used to combine clauses, resulting in complex clauses, include the following:

Time: *when, while, since, before, after, until, once*

Place: *where, wherever*

Cause: *because, since, as, now that, in as much as*

Condition: *if, unless, on condition that*

Contrast/Concession: *although, even though, despite, in spite of, while, where, whereas*

Other: *that, which, who, whoever, whom, what, why, how*

You can also point out that relative clauses—such as the bold text in the sentence, "*the earthquake **that destroyed the city** . . .*"—modify nouns, reminding students that nouns represent persons, places, things, and ideas. In this sentence, the relative clause modifies the word *earthquake*. To teach these kinds of relative clauses, ask students to circle the noun that the relative clause modifies and to underline the relative clause.

Verb Tense. Students often miss subtle shifts in verb tense that convey meaning. In the following sentences, the author switches verb tense to show a different time: ***The experiment works well for ideal gases. Some errors were made when . . .*** In presenting verb tenses, write sentences that contain different verb tenses on the board and ask students to identify the verbs (words used to show that an action is taking place or to indicate the existence of a state or condition). Then, discuss the reasons the author shifts verb tense, i.e., moves between the present, past, and future.

Passive Structures. Teaching passive structures need not be difficult. You can identify several passive structures used in the reading and can write them on the board. Then, ask students to identify the true subjects of the sentences. Explain that the subject is the part of the sentence that contains a noun or noun phrase, that is the topic of the rest of the sentence, and that agrees with the verb. One function of passive structures in science text is to make the author seem objective and academic. Another function is to make the subject of the sentence less culpable. For instance, consider the following sentence: ***The trees were eliminated.*** In this sentence we don't know who is responsible for the trees' elimination because the true subject has been deleted. Did Americans cut down the trees, or was the responsible party the lumber industry or a specific nation?

> **Use Caution Ahead**
>
> **If your students can't identify the nouns that pronouns represent or even understand the pronouns that represent nouns, they are sure to experience reading comprehension difficulties.**

Prepositions. Reading comprehension improves with an understanding of the meanings of prepositions, such as *above, beside, by, from, near, to, up to, toward,* and *until,* in the context of the sentences in which the prepositions occur. (Remember: A preposition is a word that links a noun or pronoun to another part of a sentence. One fun device for remembering prepositions is to think of a squirrel and a tree. A squirrel can run *up, down, around, into,* and *through* the branches of a tree.)

A good way to teach prepositions is to write sentences that contain them on the board and to discuss their meanings. Another useful technique is to give students a passage from their text and to ask them to circle the prepositions and discuss their meanings with a partner. You can also use *cloze activities,* which involve taking a passage of the science text and, after leaving the first and last sentences intact, deleting the prepositions and asking your students to fill in the blanks. Dictation activities also help students learn prepositions.

Adjectives and Adverbial Forms. Adjectives and adverbial forms, such as *almost, probably, never, exactly, unless, hardly, scarcely, rarely, next, not quite, last, older, younger, most, many, less, longer, least,* and *higher,* modify the meanings of sentences and words. Point out these words to students and discuss their meanings when reviewing science readings and unpacking test questions in class. Write sentences that contain these words on the board and ask students to identify the word or words they modify.

Pronouns. If your students can't identify the nouns that pronouns represent or even understand the pronouns that represent nouns, they are sure to experience reading comprehension difficulties. To help students understand that pronouns represent nouns, write sentences containing pronouns on the board and ask students to identify the nouns that the pronouns represent. You can also copy passages from the textbook and ask students to circle the pronouns and to underline the noun referents. In addition, you can use cloze activities, deleting the pronouns and asking students to infer the pronouns from corresponding nouns.

Reading Fluency

Fluency consists of the ability to read out loud with speed, accuracy, and proper expression. The National Reading Panel recommends two approaches to teaching fluency: *guided repeated oral reading* and *silent reading*. Both types of reading call for students to read text that is written at their independent reading level, a level at which they can read comfortably with 95 percent accuracy. In guided repeated oral reading, you direct students to read passages out loud, and you provide them with systematic and explicit guidance and feedback.

To gain reading fluency, students need to read and re-read, usually four times.

Strategies for Fluency Development

 Give students plenty of opportunities to practice reading and re-reading science passages.

 Don't ask students to read text written beyond their independent reading level orally in round-robin activities.

Model reading passages aloud before asking English Language Learners to read aloud (allowing them to hear the pronunciation of words and stress patterns).

Have students repeat sentences clause by clause after you; then ask them to practice reading the passages in pairs. Have them change partners several times until they can read the passages accurately at a normal pace without hesitations.

In silent reading, require students to read articles and books that are related to their science lessons and that are written at their independent reading levels. Remember that many students lack access to appropriate materials in their homes and communities; you will need to give them scientific reading materials to take home.

In Conclusion

Nationwide, English Language Learners are performing poorly in their science classes. Their failure to master the literacy skills needed to access science texts has far wider negative consequences than simply their failure to pass national assessments. English Language Learners, in fact the majority of your students, require instruction to access their reading. Fortunately, you can provide this instruction and, in so doing, can prepare these students for high-stakes exams, high-paying jobs, and higher education. ■

Bibliography

Books

Barton, Mary Lee, and Deborah L. Jordan. 2002. *Teaching reading in science: A supplement to teaching reading in content areas.* Alexandria: ASCD. www.ascd.org.

Dobb, Fred. 2004. *Essential elements of science instruction for English learners.* Los Angeles: California Science Project. http://csmp.ucop.edu/csp.

Harvey, Stephanie. 1998. *Nonfiction matters: Reading, writing, and research in grades 3–8.* Portland, Maine: Stenhouse Publishers

Nation, I.S.P. 2001. *Learning vocabulary in another language.* Cambridge: Cambridge University Press.

Scarcella, Robin. 2003. *Accelerating academic English: A focus on the English learner.* Oakland: University of California.

Schleppegrell, M. J. 2004. *The language of schooling: A functional linguistics approach.* Mahwah, N.J.: Lawrence Erlbaum.

Articles and Research Reports

Bienlenberg, Brian, and Lily Wong Fillmore. 2004. *The English they need for the test.* Association for Supervision and Curriculum Development. http://www.barrow.k12.ga.us/esol/The_English_They_Need_for_the_Test.pdf.

Corson, D. 1997. The learning and use of academic English words. *Language Learning* 47 (4): 671–718.

Gersten, Russell, and Scott Baker. 2000. What we know about effective instructional practices for English-language learners. *Exceptional Children* 66 (4): 454–70.

Lee, Okhee. 2005. Science education and student diversity: Synthesis and research agenda. *Journal of Education for Students Placed at Risk* (JESPAR) 10 (4): 431–440.

Lee, Okhee, and Mary Evalos. 2003. Integrating science with English language development. *SEDL Letter,* December.

Merino, Barbara, and Robin Scarcella. 2005. Teaching science to English learners. *UC Linguistic Minority Research Institute Newsletter* 14 (4): 1–7.

Stoddart, Trish, America Pinal, Marcia Latzke, and Dana Canaday. 2002. Integrating inquiry science and language development for English language learners. Journal of Research in Science Teaching, October, 664–687.

Wong Fillmore, Lily, and Catherine Snow. 2002. What teachers need to know about language. In *What Teachers Need to Know about Language,* ed. Carolyn Adger, Catherine E. Snow, and Donna Christian, pp. 7-54. Washington, D.C.: Center for Applied Linguistics.

Online Resources

Center for Applied Linguistics

http://www.cal.org/admin/about.html
The Center for Applied Linguistics is dedicated to improving
the teaching and learning of English learners.

The Office of English Language Acquisition, Language Enhancement, and Academic Achievement for Limited English Proficient Students (OELA)

http://www.ed.gov/abiut/offices/list/oela/index.html
The Office of English Language Acquisition, Language
Enhancement, and Academic Achievement for Limited English
Proficient Students provides assistance in carrying out national
school reforms designed to help English learners.